GRAPHIC
Nobel

Nobel prizes in physics explained with cartoons

1901 – 1910

Manu Cornet

For my brilliant physics and chemistry teachers (in order of appearance in my personal movie): Jérôme Bonaldi, Michel Lagouge, Stéphane Mansuy, Jean-Philippe Bouchaud, Bernard Castaing and Walter Lewin.

Many thanks to Annie Chen, Nadine Ho, Sara Segel, Denise Wang and Monica Wright for reading early versions of this book and providing valuable feedback.

I loathe forewords and prefaces, and I'm sure you do too. So here we go.

1901

WILHELM RÖNTGEN

In recognition of the extraordinary services he has rendered by the discovery of the remarkable rays subsequently named after him.

"X-RAY" PROBABLY MAKES YOU THINK OF THIS:

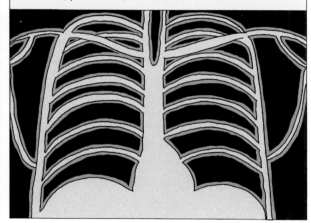

WELL, WE ALL HAVE THIS GRUMPY LOOKING 19/20th CENTURY GERMAN GUY TO THANK FOR THOSE RAYS:

ACH.

NOW TRY TO PICTURE YOURSELF BACK IN 1895...

THE FIRST CARS WERE BEING COMMERCIALIZED.

THE FIRST MOVIES WERE BEING SCREENED IN PUBLIC.

OSCAR WILDE WAS CONVICTED BECAUSE OF HIS HOMOSEXUALITY.

TELEPHONES WERE STARTING TO SPREAD.

AND HERE IS OUR MR RÖNTGEN SLAVING AWAY IN HIS LABORATORY ON A FRIDAY NIGHT (NOV 8th 1895).

HE WASN'T LOOKING FOR X-RAYS AT ALL, BUT HE WAS PLAYING WITH A DEVICE THAT COULD SHOOT ELECTRONS LIKE A CANNON. AND HE HAPPENED TO POINT IT AT A PIECE OF METAL.

(THE ACTUAL DEVICE MIGHT HAVE LOOKED A LITTLE MORE BORING)

OF COURSE THE METAL IS MADE OF ATOMS. REMEMBER WHAT THOSE LOOK LIKE?

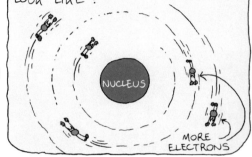

SO WE HAVE A LARGE BUNCH OF SEDENTARY ELECTRONS BEING BOMBARDED BY A STREAM OF HIGH-SPEED ELECTRONS. MANY COLLISIONS HAPPEN AND ELECTRONS GET KNOCKED OFF.

WELL, AND NOW WE HAVE... A HOLE.

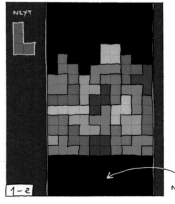

IF YOU'VE EVER PLAYED TETRIS, YOU KNOW THAT SUCH A HOLE WILL NOT LAST VERY LONG...

NOTHING

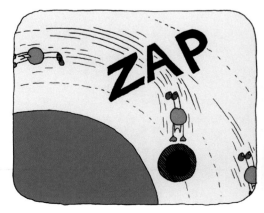

IN AN ATOM, THE FURTHER AWAY ELECTRONS ARE FROM THE NUCLEUS, THE HIGHER THEIR ENERGY.

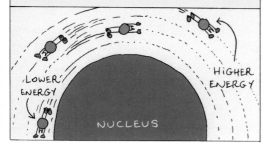

IT'S A BIT LIKE LIFTING A ROCK FROM THE GROUND. IF YOU LIFT IT HIGHER, IT HAS MORE POTENTIAL ENERGY IT CAN RELEASE WHEN YOU DROP IT AGAIN.

SO IF AN ELECTRON FROM A LOWER ENERGY RUNG IS KICKED OUT AND ITS SPOT IS TAKEN BY A HIGHER ELECTRON, THAT ELECTRON LOSES ENERGY.

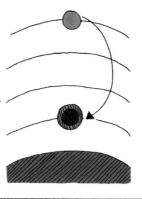

BUT WHAT DOES THAT ENERGY BECOME? PLEASE MEET MR PHOTON.

A PHOTON IS JUST A TINY CHUNK OF LIGHT. AND THEY COME IN A WHOLE RANGE OF FLAVORS.

YOUR EYES CAN SEE THESE

EACH FLAVOR IS CHARACTERIZED (AMONG OTHER THINGS) BY HOW MUCH ENERGY IT HAS. IN OUR METAPHOR HERE, LONGER LEGS REPRESENT MORE ENERGY.

LOW ENERGY

HIGH ENERGY

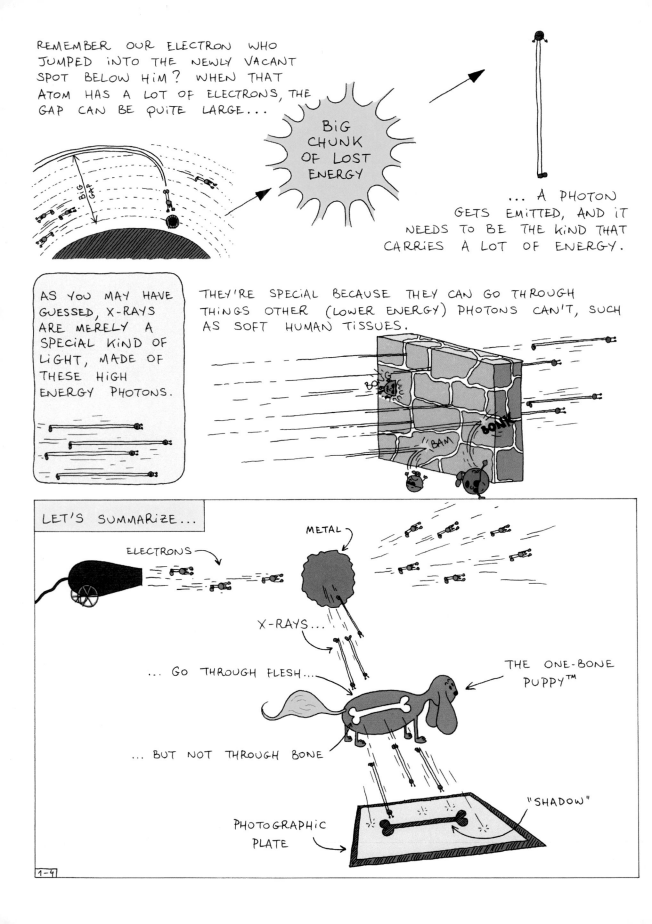

REMEMBER OUR ELECTRON WHO JUMPED INTO THE NEWLY VACANT SPOT BELOW HIM? WHEN THAT ATOM HAS A LOT OF ELECTRONS, THE GAP CAN BE QUITE LARGE...

BIG CHUNK OF LOST ENERGY

... A PHOTON GETS EMITTED, AND IT NEEDS TO BE THE KIND THAT CARRIES A LOT OF ENERGY.

AS YOU MAY HAVE GUESSED, X-RAYS ARE MERELY A SPECIAL KIND OF LIGHT, MADE OF THESE HIGH ENERGY PHOTONS.

THEY'RE SPECIAL BECAUSE THEY CAN GO THROUGH THINGS OTHER (LOWER ENERGY) PHOTONS CAN'T, SUCH AS SOFT HUMAN TISSUES.

LET'S SUMMARIZE...

ELECTRONS

METAL

X-RAYS...

... GO THROUGH FLESH...

THE ONE-BONE PUPPY™

... BUT NOT THROUGH BONE

"SHADOW"

PHOTOGRAPHIC PLATE

IT IS WORTH NOTING THAT OUR DEAR MR RÖNTGEN (AND OTHER PHYSICISTS AT THAT TIME) DIDN'T KNOW MUCH ABOUT THE ATOMIC STRUCTURE, ELECTRONS TRANSITIONING BETWEEN ENERGY LEVELS, ETC. HE MERELY STUMBLED UPON THIS UNKNOWN TYPE OF RAYS WITH INTERESTING PROPERTIES.

BUT HE VERY QUICKLY SAW THE POTENTIAL MEDICAL APPLICATIONS OF HIS DISCOVERY.

AND HE WAS WISE ENOUGH TO PROTECT HIMSELF FROM THIS LITTLE-KNOWN PHENOMENON.

THE VERY FIRST HUMAN X-RAY WAS OF RÖNTGEN'S WIFE'S HAND.

JA, OVER THERE.

NONO, NOT DANGEROUS.

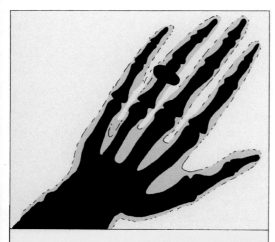

SHE SAID: "I HAVE SEEN MY DEATH."

WILHELM RÖNTGEN WANTED HUMANKIND TO IMMEDIATELY BENEFIT FROM HIS DISCOVERY AND ITS APPLICATIONS, AND HE REFUSED TO TAKE OUT PATENTS.

PIERRE AND MARIE CURIE WOULD DO THE SAME SEVERAL YEARS LATER. BUT THAT'S ANOTHER STORY.

ELECTRON

1902

HENDRIK LORENTZ & PIETER ZEEMAN

In recognition of the extraordinary service they rendered by their researches into the influence of magnetism upon radiation phenomena.

PHYSICISTS SOMETIMES LIKE TO ASK WEIRD QUESTIONS.

BACK IN THE LATE XIXth CENTURY, THE GOOD Mr LORENTZ...

HALLO!

... HAD A THEORY ABOUT THAT.

HE BELIEVED THAT LIGHT IS PRODUCED BY THE RAPID VIBRATION OF MICROSCOPIC PARTICLES.

HEEEEEEEYY

LATER SCIENTISTS HAVE DETERMINED THAT THOSE PARTICLES ARE USUALLY ELECTRONS MOVING AROUND A NUCLEUS.

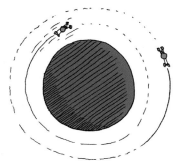

IF WE ZOOM IN, WE CAN BETTER SEE THE DIFFERENT ENERGY LEVELS. THE FARTHER AWAY FROM THE NUCLEUS, THE HIGHER THE ENERGY.

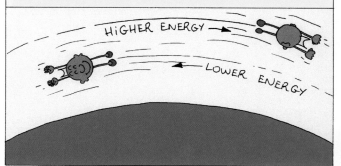

HIGHER ENERGY

LOWER ENERGY

BUT THE ELECTRONS CAN ONLY EXIST AT CERTAIN PREDEFINED LEVELS AND NOTHING IN BETWEEN. A BIT LIKE A TINY CAT ON A LARGE STAIRCASE.

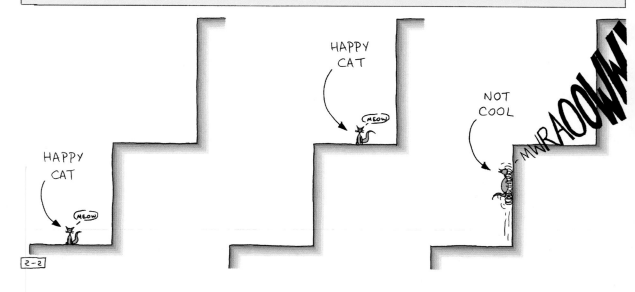

HAPPY CAT

MEOW

HAPPY CAT

MEOW

NOT COOL

MWRAOOWW

HAPPY CAT

MEOW

LET'S REPRESENT THIS IN A SIMPLISTIC WAY:

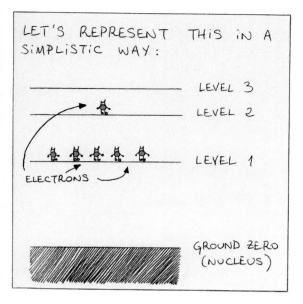

IF YOU FIND A WAY TO GIVE THESE ELECTRONS A BIG BUNCH OF ENERGY (WHICH IS EASY TO DO), YOU'LL SEE THEM JUMP TO HIGHER LEVELS.

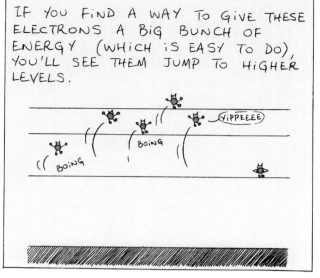

BUT AFTER A SHORT WHILE, THEY'LL GET BORED UP THERE AND WILL WANT TO GO BACK HOME. THAT MEANS THEY WILL RELEASE ENERGY AS THEY JUMP BACK DOWN.

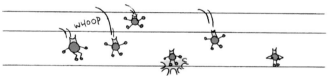

AND AS YOU MIGHT REMEMBER FROM THE PREVIOUS CHAPTER, THAT ENERGY IS RELEASED IN THE SHAPE OF A PHOTON, A TINY PACK OF LIGHT.

HOW LONG ARE ITS LEGS (THAT'S ALWAYS THE RELEVANT QUESTION WITH PHOTONS)? IN OTHER WORDS, HOW MUCH ENERGY DOES IT PACK? THAT DEPENDS ON THE HEIGHT OF THE CORRESPONDING ELECTRON'S JUMP.

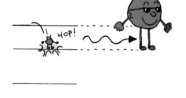
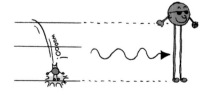
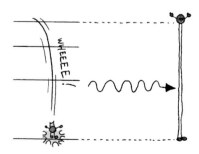

TO SEE THIS PHENOMENON IN PRACTICE, WE NEED TO GATHER A BUNCH OF IDENTICAL ATOMS (SO THAT THE PREDEFINED LEVELS FOR ELECTRONS ARE ALL THE SAME) AND PUMP SOME ENERGY INTO THAT MATERIAL TO MAKE ELECTRONS JUMP UP AND DOWN.

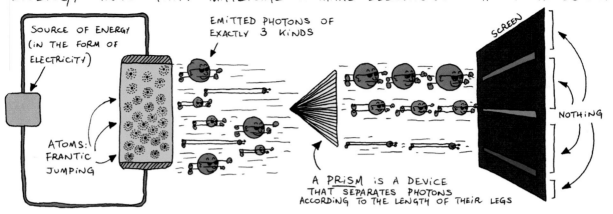

NOTE THAT WE SEE ▮▮▮▮ (THAT IS CALLED A SPECTRUM) AND NOT ▮▮▮ AS WOULD BE THE CASE IF ALL SORTS OF PHOTONS WERE EMITTED.

WHAT HE FOUND OUT WAS THAT WHEN THE MAGNET WAS CLOSE ENOUGH TO THE SOURCE OF THE LIGHT, THE BRIGHT LINES ON THE SCREEN WOULD START TO SPLIT.

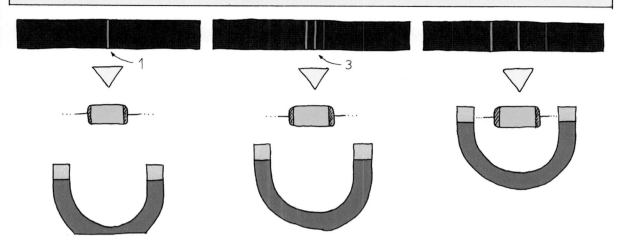

LORENTZ (WHO USED TO BE ZEEMAN'S PROFESSOR) HEARD ABOUT THESE FINDINGS ON FRIDAY 21 OCT 1896...

... AND BY THE FOLLOWING MONDAY HE HAD DRAFTED A THEORETICAL EXPLANATION THAT HE PRESENTED TO ZEEMAN.

HOW MUCH TIME LORENTZ SPENT WITH HIS FAMILY DURING THAT WEEKEND IS ANYONE'S GUESS.

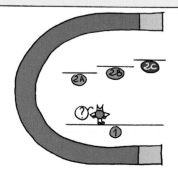

GIVEN THAT CONCEPTS SUCH AS QUANTIZED ENERGY LEVELS OR EVEN ELECTRONS WERE STILL UNKNOWN, LORENTZ'S THEORY IS NOW QUITE OUTDATED, BUT HERE IS WHAT MORE MODERN PHYSICS HAVE TO SAY ABOUT THIS.

IT TURNS OUT THAT A MAGNETIC FIELD (SUCH AS THE ONE CREATED BY THE BIG MAGNET) CAN SPLIT SOME OF THESE LEVELS INTO SEVERAL SUB-LEVELS.

WHEN THAT HAPPENS, JUMPY ELECTRONS HAVE THAT MANY MORE PLACES TO HOP UP TO, AND DOWN FROM.

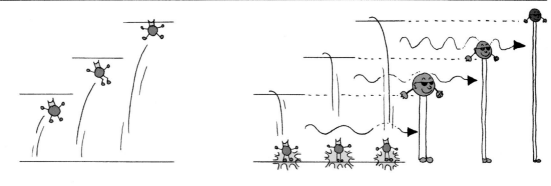

AND WHERE THERE USED TO BE A SINGLE LINE ON THE SCREEN, THERE ARE NOW SEVERAL.

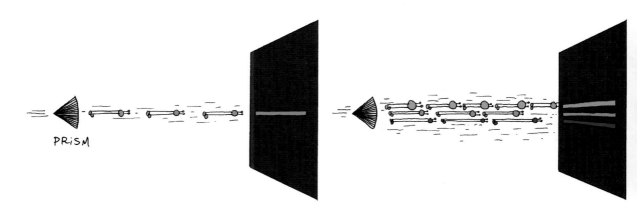

PRISM

THE BAD NEWS IS THAT ZEEMAN WAS FIRED FOR DISOBEYING A DIRECT ORDER.

IMAGINE THAT HAPPENING TO YOU; HOW WOULD YOU LIKE, AS REVENGE...

2-6

...TO GET THE NOBEL PRIZE FOR PHYSICS A FEW YEARS LATER (ALONG WITH YOUR FORMER PROFESSOR)?

ONE LAUREATE IN 1901, TWO IN 1902, WILL THERE BE THREE IN 1903? THAT'S ANOTHER STORY.

1903

HENRI BECQUEREL, PIERRE CURIE & MARIE SKŁODOWSKA CURIE

In recognition of the extra-ordinary services he has rendered by his discovery of spontaneous radioactivity.

In recognition of the extra-ordinary services they have rendered by their joint researches on the radiation phenomena discovered by Professor Henri Becquerel.

FAMILY GATHERINGS MUST HAVE BEEN FUN AT THE BECQUERELS'.

HENRI (AGE 26) PHYSICIST

ÉDOUARD (AGE 58) PHYSICIST

ANTOINE (AGE 89) PHYSICIST

JEAN (AGE -0.5) FUTURE PHYSICIST

AND THEN HE SAID: JUST ADD A PINCH OF POTASSIUM!!! HA HAHAHAHA HAHAHAH HAHA

CHRISTMAS 1877.

LET'S FOCUS ON THE GUY ON THE LEFT, AND FAST-FORWARD A COUPLE DECADES (1896).

HENRI BECQUEREL (AGE 44)

RÖNTGEN'S 1895 X-RAY DISCOVERY GOT HIM ALL EXCITED.

HENRI, ALL EXCITED

HE STARTED EXPOSING OTHER MATERIALS TO SUNLIGHT, TO SEE IF THEY, TOO, WOULD EMIT ANYTHING LIKE X-RAYS.

URANIUM COMPOUND

MALTESE CROSS

PHOTOGRAPHIC PLATE

3-1

UNFORTUNATELY, AS M. BECQUEREL WAS TRYING TO REPRODUCE HIS EXPERIMENT, THE WEATHER IN PARIS WAS TERRIBLE...

SACREBLEU.

... SO HE KEPT HIS URANIUM IN A DRAWER.

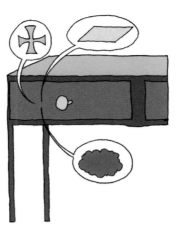

BUT WHEN HE TOOK IT OUT AGAIN, LO AND BEHOLD, THE SAMPLE HAD IMPRESSED THE PHOTOGRAPHIC PLATE, WITHOUT ANY SUNLIGHT.

PALSAMBLEU.

BECQUEREL HAD OBSERVED WHAT IS CALLED "SPONTANEOUS RADIOACTIVITY". IT WAS RATHER UNEXPECTED AT THE TIME THAT SOMETHING WOULD EMIT SOME SORT OF LIGHT, OR RAYS, WITHOUT HAVING BEEN "EXCITED" OR OTHERWISE TAMPERED WITH, THUS RELEASING ENERGY SEEMINGLY OUT OF NOWHERE. HOW COULD THAT BE POSSIBLE?

IT TURNS OUT NATURE IS A BIT OF A LAZY BUM.

HMMPH!

IT'S ALWAYS TRYING TO LET THINGS DROP TO THEIR STATE OF LOWEST ENERGY.

OR IN OTHER WORDS, THEIR MOST STABLE STATE.

ROLROLROLROLROLROLLL

BUT WHAT HAPPENS IF THE PATH TO THE MOST STABLE STATE IS BLOCKED?

THINGS REALLY WANT TO DROP DOWN TO THAT STATE, AND THEY WILL, EVENTUALLY. BUT IT MIGHT TAKE A WHILE.

OR, A LITTLE BIT OF ENERGY MAY NEED TO BE SPENT FIRST, BEFORE A WHOLE LOT OF ENERGY CAN BE FREED UP.

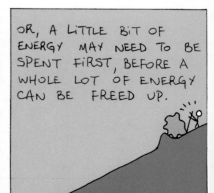

ANOTHER EXAMPLE: YOU, ON YOUR SOFA. IT'S LATE, YOU'RE FALLING ASLEEP; YOU'D BE MUCH MORE COMFORTABLE IN YOUR BED...

MRRRRF

... IT TAKES A BIT OF ENERGY FOR YOU TO GET UP...

GRMBL GRMBL

... AND EVENTUALLY REACH THIS MORE STABLE STATE FOR THE NIGHT.

YOU CAN CONSIDER A PIECE OF RADIOACTIVE MATERIAL AS A BUNCH OF ATOMS THAT HAVE FALLEN ASLEEP ON THE SOFA.

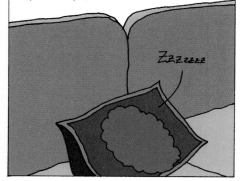

ZZZZZZZ

TO UNDERSTAND WHY, WE NOW NEED TO FOCUS ON THE ATOM'S NUCLEUS INSTEAD OF ITS ELECTRONS.

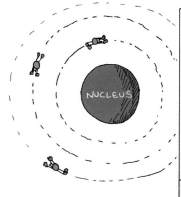

NUCLEUS

A LITTLE CLOSER...

THERE WE GO.

PROTON

NEUTRON

NUCLEI ARE MEANT TO HAVE (VERY ROUGHLY) THE SAME NUMBER OF PROTONS AND NEUTRONS, AND NOT TO BE TOO HUGE.

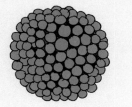

THAT'S WHEN THEY ARE MOST STABLE.

BUT IT CAN HAPPEN THAT THEY HAVE...

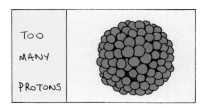

TOO MANY PROTONS

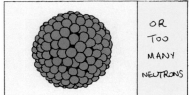

OR TOO MANY NEUTRONS

OR THAT THEY'RE JUST TOO BIG ALTOGETHER.

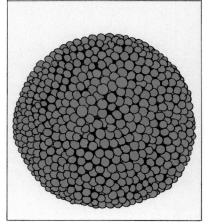

IF A NUCLEUS IS UNSTABLE, IT WILL SPONTANEOUSLY TRY TO "FIX" ITSELF.

HERE ARE 3 COMMON WAYS FOR IT TO DO SO. ALL 3 ARE CALLED "RADIOACTIVITY".

α (ALPHA)　　β (BETA)　　γ (GAMMA)

TO GET SMALLER, IT CAN SPIT OUT 2 PROTONS AND 2 NEUTRONS. THIS IS CALLED "ALPHA DECAY".

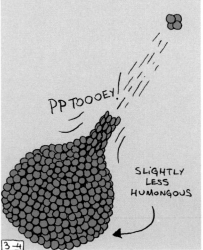

PPTOOOEY!

SLIGHTLY LESS HUMONGOUS

3-4

IF IT HAS TOO MANY NEUTRONS, IT CAN TRANSFORM ONE OF THEM INTO ONE PROTON + ONE ELECTRON, AND SPIT OUT THE ELECTRON.

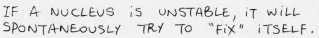

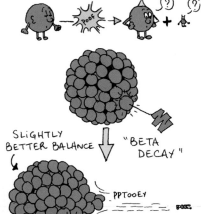

SLIGHTLY BETTER BALANCE

"BETA DECAY"

PPTOOEY

OFTEN, AFTER EITHER "α" OR "β" DECAY, THE NUCLEUS WILL STILL BE IN A BIT OF AN EXCITED STATE.

TWITCH

IT WILL THEN CALM BACK DOWN BY LETTING OUT SOME ENERGY = A PHOTON.

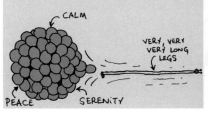

CALM

VERY, VERY VERY LONG LEGS

PEACE　SERENITY

IT IS WORTH NOTING THAT THESE "DECAY" EVENTS CAN HAPPEN IN SE-
QUENCE, UNTIL A MORE STABLE STATE IS REACHED.

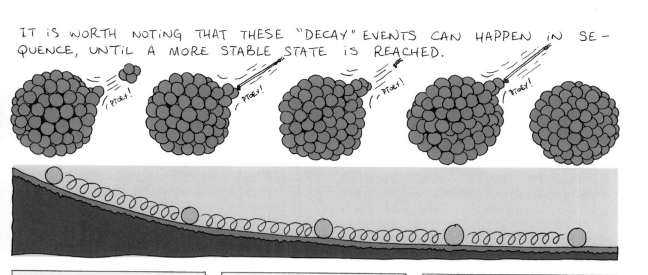

IF RADIOACTIVE ATOMS ARE SO UNSTABLE, YOU MIGHT WONDER, SURELY THERE SHOULDN'T BE ANY LEFT BY NOW? BUT REMEMBER: THEY, TOO, ARE ASLEEP ON THE SOFA...

... AND THEY ARE VERY, VERY LAZY.

ALL RIGHT, I'LL GO TO BED.

IN FIVE MILLION YEARS.

BUT SINCE THERE ARE SO MANY ATOMS IN A SMALL CHUNK OF RADIO-ACTIVE MATERIAL, A CONS-TANT FLOW OF PARTICLES IS EMITTED.

AND THAT WAS MORE THAN ENOUGH TO IM-PRESS M. BECQUEREL'S PHOTOGRAPHIC PLATE.

ALSO INTERESTED IN THIS TOPIC WERE MARIE CURIE AND HER HUSBAND PIERRE WHO, IN SEARCH OF OTHER RADIOACTIVE MATERIALS, DISCOVERED TWO NEW CHEMICAL ELEMENTS.

WHAT'S UP WITH MALE PHYSICISTS AND BEARDS? →

UNFORTUNATELY, THEY WERE NOT AS DILIGENT AS RÖNT-GEN IN PROTECTING THEMSELVES...

... AND MARIE WOULD EVENTUALLY DIE OF THE EFFECTS OF PRO-LONGED RADIATION EXPO-SURE, AS WOULD HER DAUGHTER AND SON-IN-LAW, BOTH PHYSICISTS.

AS FOR PIERRE, HE WAS RUN OVER BY A CAR A MERE 10 METERS AWAY FROM WHERE THE AUTHOR OF THIS BOOK SPENT HIS FIRST QUARTER CENTURY.

THE CURIES ALSO REFUSED TO TAKE OUT PATENTS.

TO THIS DAY, ALL OF THEIR LABORATORY BOOKS ARE TOO DANGEROUS TO TOUCH. EVEN MARIE'S COOKBOOKS. BUT THAT'S ANOTHER STORY.

3-5

NUCLEUS

1904

LORD RAYLEIGH (a.k.a. JOHN WILLIAM STRUTT)

For his investigations of the densities of the most important gases and for his discovery of argon in connection with these studies.

YOU'VE LIKELY SEEN THIS BEFORE:

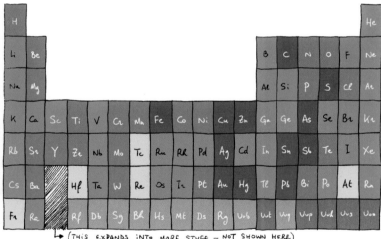

(THIS EXPANDS INTO MORE STUFF — NOT SHOWN HERE)

THE COLORS REPRESENT THE PERIOD WHEN EACH ELEMENT WAS DISCOVERED. BY 1894 (THE YEAR WE ARE INTERESTED IN TODAY), A LOT OF GROUND HAD ALREADY BEEN COVERED... WELL, EXCEPT FOR THAT RIGHTMOST COLUMN HERE.

ERRM, OOH-KAAY... BUT WHY SHOULD I CARE??

ONE REASON MIGHT BE ARGON (Ar) BECAUSE YOU BREATHE IN ABOUT 100 LITERS (25 GALLONS) OF IT EVERYDAY.

PWRFWHAT

WE'LL GET BACK TO THAT BUT DON'T WORRY, IT'S NOT TOXIC. IN FACT, THE ELEMENTS IN THAT RIGHT-MOST COLUMN ARE THE SNOBBIEST OF ALL: THEY DON'T INTERMINGLE WITH OTHERS.

ME? WITH THE PLEBS? PWAAH!

THAT IS WHY IT WAS SO DIFFICULT TO DISCOVER THEM: THEY STAY PUT, IN THEIR LITTLE CORNER...

WELL, AT LEAST HE'S VARYING THE BEARD STYLE

... UNTIL LORD RAYLEIGH CORNERED ONE OF THEM.

4-1

LIKE MANY SCIENTISTS, LORD RAY-
LEIGH WAS INTERESTED IN ATOMS,
AND MORE GENERALLY IN WHAT
STUFF IS MADE OF.

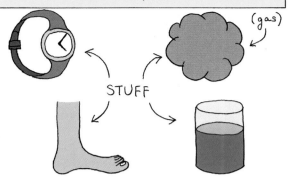

IS THERE AN ELEMENTARY BUILDING
BLOCK OF THE WHOLE UNIVERSE?
BACK IN THE XIXᵗʰ CENTURY, PEOPLE
THOUGHT THAS MIGHT BE HYDROGEN.

THIS WAS CALLED THE PROUT HYPOTHESIS. IT CAME FROM SCIENTISTS'
OBSERVATION THAT ALL ATOMS KNOWN BACK THEN SEEMED TO WEIGH A
WHOLE MULTIPLE OF THE HYDROGEN ATOM'S WEIGHT.

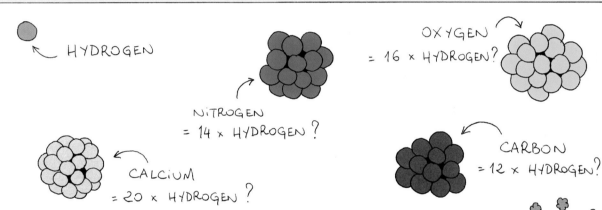

HYDROGEN

NITROGEN
= 14 × HYDROGEN?

OXYGEN
= 16 × HYDROGEN?

CALCIUM
= 20 × HYDROGEN?

CARBON
= 12 × HYDROGEN?

IF YOU THINK ABOUT IT, SINCE ELECTRONS ARE SO SMALL
AND WEIGH NEXT TO NOTHING COMPARED TO THE ATOM'S NUCLEUS,
IT MAKES SOME SENSE THAT THEY ARE EASY TO COMPLETELY IGNORE
IN THIS MODEL.

IN ORDER TO TEST THAT HYPOTHESIS,
ONE MUST VERIFY THAT NITROGEN,
FOR INSTANCE, WEIGHS EXACTLY
14 TIMES MORE THAN HYDROGEN,
AND NOT 13.8 OR 14.5, OTHER-
WISE THE MODEL IS WORTHLESS. GOOD
PRECISION IS MANDATORY.

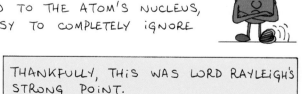

THANKFULLY, THIS WAS LORD RAYLEIGH'S
STRONG POINT.

PRRRRRECISION
IS PARRRAMOUNT.

HALF A
HYDROGEN??

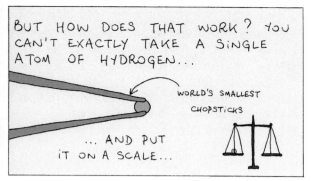

BUT HOW DOES THAT WORK? YOU CAN'T EXACTLY TAKE A SINGLE ATOM OF HYDROGEN...

WORLD'S SMALLEST CHOPSTICKS

... AND PUT IT ON A SCALE...

... AND IF YOU TAKE A BUNCH OF NITROGEN...

LIQUID NITROGEN (VERY COLD)

... HOW DO YOU KNOW HOW MANY ATOMS THERE ARE?

ONE TRICK THAT LORD RAYLEIGH USED WAS TO WORK WITH GASES. SEE, GASES HAVE A VERY CONVENIENT PROPERTY:

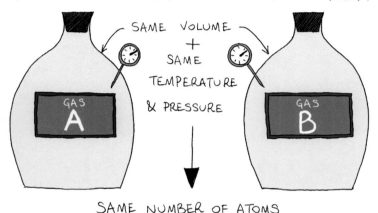

SAME VOLUME + SAME TEMPERATURE & PRESSURE

GAS A

GAS B

SAME NUMBER OF ATOMS

IF YOU TAKE THE SAME VOLUME OF TWO DIFFERENT GASES A AND B AND YOU MAKE SURE THEIR TEMPERATURE AND PRESSURE ARE THE SAME THEN YOU KNOW THE NUMBER OF ATOMS OF A AND B ARE ALSO THE SAME. THIS ALSO WORKS WITH NUMBERS OF MOLECULES, FOR EXAMPLE O_2 (OXYGEN GAS), CO_2 (CARBON DIOXIDE) OR H_2 (HYDROGEN GAS).

IN PRACTICE, IF YOU CAN GET YOURSELF TWO CONTAINERS WITH THE SAME VOLUME, TEMPERATURE AND PRESSURE, ONE FILLED WITH H_2 AND THE OTHER WITH N_2, AND IF YOU CAN SHOW THAT THE SECOND ONE WEIGHS EXACTLY 14 TIMES MORE THAN THE FIRST, YOU CAN DEDUCE THAT EACH N_2 MOLECULE WEIGHS THE SAME AS 14 H_2 MOLECULES, AND THEREFORE THAT EACH NITROGEN ATOM WEIGHS THE SAME AS 14 HYDROGEN ATOMS. THAT'S A GOOD DATA POINT FOR TRYING TO VERIFY THE PROUT HYPOTHESIS.

SOOO... WHAT DID THE POOR LORD RAYLEIGH DO EVERYDAY? GUESS...

ME?

OOOH...

I'M WEIGHING NITROGEN.

AGAIN.

BACK THEN IN 1894, IT WAS ALREADY KNOWN THAT AIR WAS MADE OF APPROXIMATELY:

OXYGEN (O_2) 20%

NITROGEN (N_2) 80%

AIR =

RAYLEIGH THUS GOT HIMSELF SOME AIR AND USED SOME COMMON TECHNIQUES FOR GETTING RID OF THE O_2 PART (FOR INSTANCE: BURN SOME STUFF UNTIL IT NO LONGER WANTS TO BURN), UNTIL HE WAS LEFT WITH...

OH GREAT.

MORE NITROGEN.

JUST WHAT I NEEDED.

BUT HERE'S THE CATCH. RAYLEIGH ALSO KNEW HOW TO SYNTHESIZE N_2 USING OTHER CHEMICAL MEANS (NOT FROM AIR). AND THE N_2 HE EXTRACTED FROM AIR WAS ALWAYS A LITTLE BIT HEAVIER. WERE THOSE NOT THE SAME GAS?

FRUSTRATION

PUZZLEMENT

UNCERTAINTY

BEARD

DOUBT

RAYLEIGH GAVE A LECTURE ABOUT HIS RESEARCH ON THURSDAY 19 APR. 1894.

LUCKILY, IN THE AUDIENCE WAS ONE WILLIAM RAMSAY.

OKAY NOW THIS IS GETTING OLD

RAYLEIGH AND RAMSAY STARTED WORKING TOGETHER AND CONCLUDED THAT AIR MUST CONTAIN ANOTHER, YET UNKNOWN, COMPONENT, HEAVIER THAN NITROGEN.

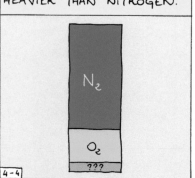

4-4

THE BULK OF THEIR WORK WAS THEN TO SEPARATE N_2 FROM THAT UNKNOWN COMPONENT.

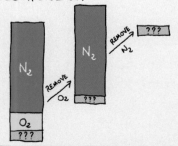

THEY SUCCEEDED AND CALLED IT ARGON, FROM THE GREEK "LAZY" OR "INACTIVE".

AIR IS ABOUT 1% ARGON. THAT'S WHY YOU BREA-THE IN SO MUCH OF IT.

OTHER STUFF (WATER, CO_2, ETC.)

THE DISCOVERY OF ARGON LED TO THAT OF OTHER "INERT GASES", IN THE SAME COLUMN OF THE PERIODIC TABLE, SUCH AS NEON, KRYPTON OR XENON.

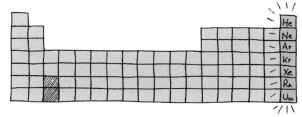

AT THIS POINT, IT IS PERHAPS WORTH NOTING THAT EVEN AFTER JUST A FEW YEARS OF NOBEL PRIZES, WE CAN ALREADY SEE THE IMPORTANCE OF COLLABORATION...

...AS WELL AS THE FREQUENT OCCURRENCE OF DISCOVERIES THAT ARE ACTUALLY QUITE DIFFERENT FROM THE ORIGINAL GOAL OF THE RESEARCH.

SPEAKING OF INITIAL RESEARCH GOALS, WHAT ABOUT THE PROUT HYPOTHESIS? IS IT ACTUALLY CORRECT? NOT QUITE. IT TURNS OUT THAT WHEN YOU COMBINE TWO HYDROGEN ATOMS, WITH A WEIGHT OF 1 EACH (THAT'S CALLED "FUSION"), YOU DON'T GET AN ATOM OF WEIGHT 2, BUT A LITTLE BIT LESS.

1   +   1   →   < 2   +

THE DIFFERENCE BECOMES THE VERY ENERGY THAT MAKES THE SUN SHINE. BUT THAT'S ANOTHER STORY.

HIS LORDSHIP SIR ARGON, KNIGHT OF
THE NOBLE GASES, PhD

1905

PHILIPP EDUARD ANTON VON LENARD
For his work on cathode rays.

THE NOBEL COMMITEE IS A VERY SECRETIVE BUNCH.

REALLY, THE ONLY THING THAT THEY DISCLOSE IS THE ACTUAL LIST OF WINNERS.

EVERYTHING ELSE (NOMINATIONS, DEBATE, PROCESS) IS KEPT CONFIDENTIAL.

HOWEVER, SOME OF THIS INFORMATION IS MADE AVAILABLE TO THE PUBLIC, SEVERAL DECADES LATER. ONE INTERESTING BIT IS THE LIST OF NOMINATIONS.

WE NOW KNOW THAT IN 1901, RÖNTGEN GOT 10 NOMINATIONS OUT OF A TOTAL OF 35, BY 30 COMMITTEE MEMBERS. THE RUNNER-UP, WITH 6 NOMINATIONS, WAS ANOTHER GERMAN PHYSICIST NAMED PHILIPP LENARD. 5 MEMBERS OF THE COMMITTEE ACTUALLY NOMINATED RÖNTGEN AND LENARD AS A PAIR.

5-1

REMEMBER HOW RÖNTGEN HAD USED A DEVICE THAT COULD SHOOT OUT ELECTRONS?

ACTUALLY, THIS WOULD HAVE BEEN A MORE ACCURATE REPRESENTATION:

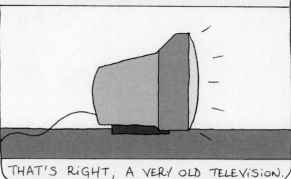

THAT'S RIGHT, A VERY OLD TELEVISION.

YOU MIGHT HAVE HEARD OF THOSE AS "CRT SCREENS" BECAUSE THE CORE TECHNOLOGY THEY USED WAS THE CATHODE RAY TUBE. THOSE WERE INVENTED AROUND 1870, AND STARTED BEING USED IN COMMERCIAL TVs AROUND 1930.

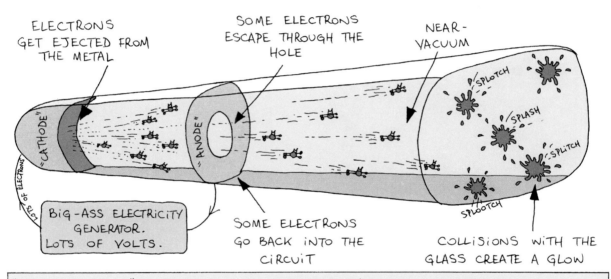

ELECTRONS GET EJECTED FROM THE METAL

SOME ELECTRONS ESCAPE THROUGH THE HOLE

NEAR-VACUUM

"CATHODE"

"ANODE"

LOTS OF ELECTRONS

BIG-ASS ELECTRICITY GENERATOR. LOTS OF VOLTS.

SOME ELECTRONS GO BACK INTO THE CIRCUIT

SPLOTCH
SPLASH
SPLITCH
SPLOOTCH

COLLISIONS WITH THE GLASS CREATE A GLOW

"CATHODE RAYS" ARE WHAT WE TODAY CALL "ELECTRONS". SAME THING.

LENARD'S CONTRIBUTION WAS THE INVENTION (AROUND 1892) OF "LENARD WINDOWS".

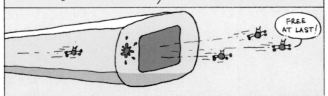

FREE AT LAST!

MADE OF VERY THIN METAL, THEY MAINTAINED A TIGHT SEAL TO PRESERVE THE VACUUM, BUT THEY ALLOWED SOME ELECTRONS TO ESCAPE OUT, THUS MAKING THEM A LOT EASIER TO STUDY.

5-2

WE'RE NOT SURE WHETHER RÖNTGEN ACTUALLY USED A TUBE ENHANCED WITH LENARD'S INVENTION...

... BUT IT'S ENTIRELY POSSIBLE, AND WOULD HELP EXPLAIN LENARD'S NOMINATIONS IN 1901.

BUT LENARD DIDN'T GET IT.

AND HE WAS A VERY SORE LOSER.

HE LATER WROTE A LIST OF REASONS WHY HE HADN'T DISCOVERED X-RAYS DESPITE HAVING HAD ACCESS TO THE SAME DEVICES RÖNTGEN HAD.

BUT THE TRUTH IS THAT HE MADE THE SAME MISTAKE AS MANY SCIENTISTS: HE PAID MORE ATTENTION TO WHAT HE WAS LOOKING FOR THAN TO WHAT HE ACTUALLY SAW.

ONLY THE GREATEST AVOID THIS TRAP.

SO LENARD WASN'T THE ONE WHO DISCOVERED X-RAYS.

AND EVEN THOUGH HE DID GET THE NOBEL IN 1905 FOR HIS UNARGUABLY MAJOR CONTRIBUTIONS, HE WAS STILL UNSATISFIED...

... THEREBY INVENTING THE INTERESTING CONCEPT OF THE "SORE NOBEL PRIZE WINNER".

HE WAS STILL UNSATISFIED IN PART BECAUSE ANOTHER GERMAN PHYSICIST WAS STARTING TO STEAL THE LIMELIGHT THAT HIS DELUSIONAL SELF THOUGHT HE DESERVED MORE.

LENARD AND EINSTEIN HAD VERY LITTLE IN COMMON. EINSTEIN WAS A GENIUS THEORIST WHEREAS LENARD BELIEVED ONLY IN DOING EXPERIMENTS.

DROWNING IN HIS OWN JEALOUSY, LENARD STARTED CALLING EINSTEIN'S CONTRIBUTIONS "JEWISH PHYSICS". ARE YOU STARTING TO HAVE A BAD FEELING ABOUT THIS GUY?

REMEMBER, THIS WAS EARLY XXth CENTURY GERMANY...

... YUP.

5-4

LENARD JOINED THE NATIONAL SOCIALIST PARTY EARLY ON, BEFORE IT BECAME POPULAR; HE BECAME ITS "CHIEF OF ARYAN PHYSICS" AND REMAINED AN ACTIVE PROPONENT UNTIL HIS DEATH IN 1947.

YOU MIGHT KNOW THIS PARTY BETTER UNDER THE NAME "NAZI", AND FROM ITS LEADER WHO BECAME GERMANY'S CHANCELLOR IN 1933, ADOLF HITLER. BUT THAT'S ANOTHER STORY.

1906

JOSEPH JOHN THOMSON

In recognition of the great merits of his theoretical and experimental investigations on the conduction of electricity by gases.

MANY PHYSICISTS WERE PRETTY OBSESSED WITH THESE "CATHODE RAY TUBE" DEVICES...

FROM THE 1870s WHEN THEY WERE INVENTED...

... UP UNTIL THE 1930s WHEN THEY STARTED GETTING MASS-PRODUCED FOR TELEVISION SETS.

RÖNTGEN FOUND X-RAYS WITH THEM

LENARD GAVE THEM A WINDOW

BUT NOBODY KNEW WHAT THOSE "RAYS" ACTUALLY WERE, UNTIL A BRITISH PHYSICIST NAMED J.J. THOMSON CAME ALONG WITH A FEW SMART EXPERIMENTS.

HE PLACED THE TUBE IN AN ELECTRIC FIELD AND SAW THAT THE RAYS CURVED UP.

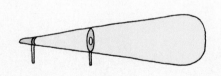

HE TOOK A PAGE FROM ZEEMAN'S BOOK AND USED A MAGNET : THE RAYS CURVED DOWN.

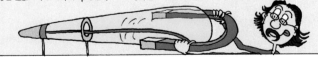

HE DID BOTH AT ONCE AND ADJUSTED THE STRENGTHS UNTIL THE RAYS WERE STRAIGHT AGAIN. THAT GAVE HIM CLUES ABOUT THE MASS AND ELECTRIC CHARGE OF WHATEVER THESE WERE.

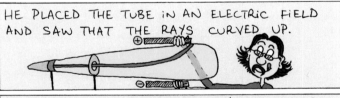

6-1

HE CONCLUDED THAT CATHODE RAYS WERE MADE OF CORPUSCLES (LATER CALLED "ELECTRONS") ABOUT 2000 TIMES LIGHTER THAN A HYDROGEN ATOM — WHICH HAD UNTIL THEN BEEN CONSIDERED THE SMALLEST BUILDING BLOCK OF ALL MATTER.

("A-TOM" COMES FROM THE GREEK "UN-CUTTABLE")

HE ALSO CONCLUDED THAT THESE CORPUSCLES MUST BE NEGATIVELY CHARGED.

FROM THESE CONCLUSIONS, HE IMAGINED A LOVELY (THOUGH INACCURATE) MODEL WHERE MATTER IS MADE OF SMALL NEGATIVE PARTICLES BATHING IN A POSITIVE MAGMA.

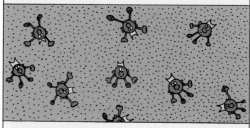

AS A GOOD ENGLISHMAN, HE CALLED IT THE PLUM PUDDING MODEL.

BESIDES BEING CREDITED FOR THE DISCOVERY OF THE ELECTRON (ALREADY NOT BAD, RIGHT?), J.J. THOMSON'S ARGUABLY EVEN BIGGER CONTRIBUTION WAS HIS EXCELLENT TEACHING AND MENTORSHIP.

THE JJ THOMSON NOBEL FACTORY

6-2

RESEARCH STUDENT UNDER JJ
PHYS. NOBEL 1917

RESEARCH STUDENT UNDER JJ
PHYS. NOBEL 1922

RESEARCH STUDENT UNDER JJ
PHYS. NOBEL 1954

RESEARCH STUDENT UNDER JJ
PHYS. NOBEL 1915

RESEARCH STUDENT UNDER JJ
PHYS. NOBEL 1928

RESEARCH STUDENT UNDER JJ
PHYS. NOBEL 1927

RESEARCH STUDENT UNDER JJ
CHEMIS. NOBEL 1922

RESEARCH STUDENT UNDER JJ
CHEMIS. NOBEL 1908

JJ'S SON
PHYS. NOBEL 1937

BUT THAT'S A WHOLE BUNCH OF OTHER STORIES.

1907

ALBERT ABRAHAM MICHELSON

For his optical precision instruments and the spectroscopic and metrological investigations carried out with their aid.

WE ARE STILL JUST A FEW YEARS BEFORE 1900. MOST SCIENTISTS BELIEVE THAT ANY PROPAGATING WAVE NEEDS A SUPPORTING MEDIUM.

SO, MISTER SCIENTIST FROM THE "BELLE ÉPOQUE," WHAT'S THE MEDIUM FOR LIGHT COMING FROM THE SUN?

AND SO THE "ETHER" WAS BORN. IT WAS POSTULATED THAT THIS THING WAS EVERYWHERE. IT HAD TO BE WEIGHTLESS AND COLORLESS AND ODORLESS SINCE NOBODY HAD EVER FELT OR SEEN OR SMELLED IT. BUT IT HAD GOT TO BE THERE, WHEREVER THERE WAS LIGHT.

THERE HAD TO BE A WAY TO DETECT IT!

IF YOU'VE EVER TRIED MAKING YOURSELF HEARD AGAINST A STRONG WIND, YOU KNOW THAT SOMETIMES THE MEDIUM DOESN'T HELP.

SINCE THE EARTH IS TURNING AROUND ITSELF PRETTY FAST, THERE MUST BE SOME MEASURABLE EFFECTS OF "ETHER WIND."

THAT'S WHAT ALBERT MICHELSON (ALONG WITH HIS COLLEAGUE EDWARD MORLEY) SET OUT TO MEASURE IN 1897.

NOW AGED 42 (AND NO LESS OBSESSED WITH THE SPEED OF LIGHT)

WHAT THEY BUILT WAS A CLEVER WAY TO SPLIT A BEAM OF LIGHT INTO TWO HALVES THAT WOULD BE AFFECTED BY THE ETHER WIND IN DIFFERENT WAYS, AND THEN GET REUNITED TO INTERFERE WITH EACH OTHER.

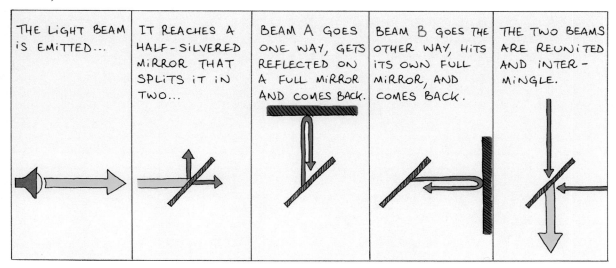

| THE LIGHT BEAM IS EMITTED... | IT REACHES A HALF-SILVERED MIRROR THAT SPLITS IT IN TWO... | BEAM A GOES ONE WAY, GETS REFLECTED ON A FULL MIRROR AND COMES BACK. | BEAM B GOES THE OTHER WAY, HITS ITS OWN FULL MIRROR, AND COMES BACK. | THE TWO BEAMS ARE REUNITED AND INTER-MINGLE. |

WHEN A LIGHT WAVE ENCOUNTERS ANOTHER, THEY CAN LEAD TO "CONSTRUCTIVE INTERFERENCE" IF THEY ARE IN PHASE WITH EACH OTHER...

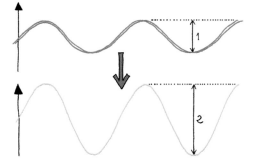

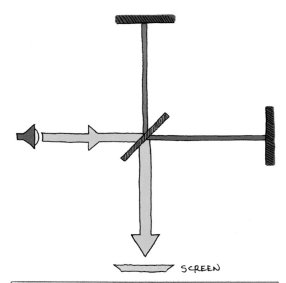

SCREEN

... AND IF THEY ARE NOT, THEY WILL LEAD TO "DESTRUCTIVE INTERFERENCE".

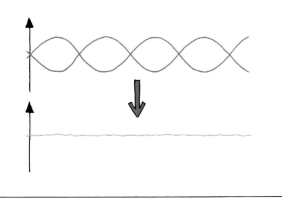

DEPENDING ON HOW THE DIRECTION OF ETHER WINDS CHANGED WITHIN A DAY, A MONTH, A YEAR, THEY WOULD AFFECT THE TWO HALF-BEAMS IN DIFFERENT WAYS, CHANGING HOW "IN PHASE" THEY WOULD BE WITH EACH OTHER. MICHELSON AND MORLEY EXPECTED THAT SUCH CHANGES WOULD IN TURN CHANGE THE INTERFERENCE PATTERNS THAT THEY SAW ON THE SCREEN.

7-3

AND THAT'S NOT EVEN THE WHOLE STORY. TO AVOID "POLLUTING" THEIR VERY PRECISE MEASUREMENTS WITH EARTHLY PHENOMENA SUCH AS VIBRATIONS COMING FROM THE GROUND, MICHELSON AND MORLEY PUT THE WHOLE THING ON A BIG SLAB OF STONE FLOATING IN LIQUID MERCURY. QUITE THE SETUP! (YUMM, THOSE POISONOUS MERCURY FUMES...)

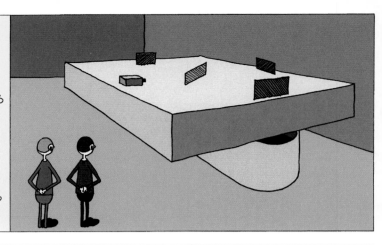

BUT THE EXPERIMENT WAS A FAILURE. MICHELSON DIDN'T FIND ANY SIGN OF ETHER WINDS WHATSOEVER...

DESPITE

ALL THE

EXPERIMENTAL

CONDITIONS

THAT

HE

TRIED.

HOWEVER, NOT ONLY DID THE FAILED EXPERIMENT HELP SHOW THAT THERE'S REALLY NO ETHER (LIGHT HAPPILY PROPAGATES IN A VACUUM), BUT THE MICHELSON INTERFEROMETER IS, TO THIS DAY, A REMARKABLY PRECISE INSTRUMENT OF MEASURE.

AS FOR HIS OBSESSION ABOUT THE SPEED OF LIGHT, MICHELSON LATER ON MEASURED IT AT 299 774 km/s, A MERE 0.006 % LOWER THAN THE COMMONLY ACCEPTED VALUE NOWADAYS OF 299 792 km/s.

AND WHAT THE EXPERIMENT HINTED AT, THE SPEED OF LIGHT AS A UNIVERSAL CONSTANT, WOULD SOON BE AT THE HEART OF EINSTEIN'S THEORY OF RELATIVITY.

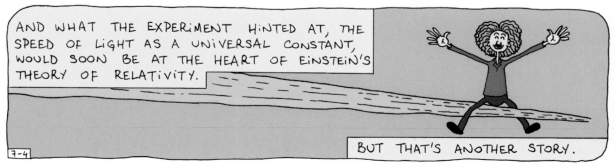

BUT THAT'S ANOTHER STORY.

1908

GABRIEL LIPPMANN

For his method of reproducing colours photographically based on the phenomenon of interference.

HAVE YOU TRIED LOOKING AT A COLOR PHOTO PRINT... REALLY CLOSELY?

CLOSER...

CLOSER...

AH! DO YOU SEE THE INDIVIDUAL DOTS?

GABRIEL LIPPMANN

BEARD

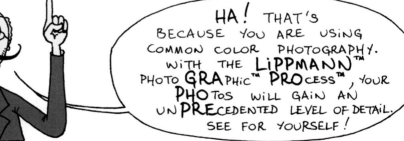

HA! THAT'S BECAUSE YOU ARE USING COMMON COLOR PHOTOGRAPHY. WITH THE LIPPMANN™ PHOTO GRAPHIC™ PROCESS™, YOUR PHOTOS WILL GAIN AN UNPRECEDENTED LEVEL OF DETAIL. SEE FOR YOURSELF!

CLOSER...   CLOSER...   CLOSER...

OUCH.

MUCH FINER GRAINED!

8-1

SO WHY THE DIFFERENCE? USUAL COLOR PHOTOGRAPHY SEPARATES THE DETECTION OF RED, GREEN & BLUE.

ON THE OTHER ~~GLOVE~~ HAND, LIPPMANN PHOTOGRAPHY RELIES ON INCOMING LIGHT INTERFERING WITH ITSELF.

TO UNDERSTAND HOW THAT WORKS, LET'S TALK ABOUT VIOLINS.

WELL, YES, HIM. BUT APART FROM THAT?

THE ANSWER IS: STANDING WAVES.

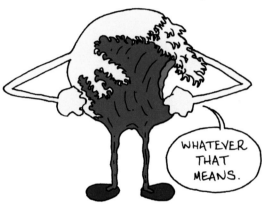

WHEN YOU BOW A VIOLIN STRING, IT VIBRATES. WAVES START TRAVELLING UP AND DOWN THE STRING.

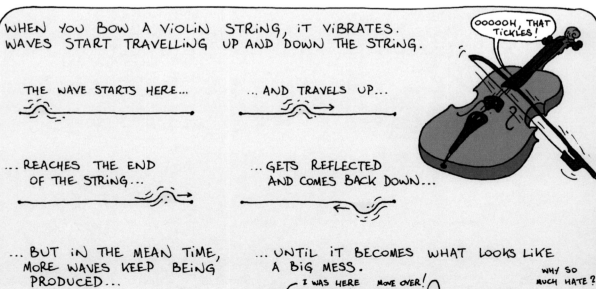

THE WAVE STARTS HERE...

...AND TRAVELS UP...

...REACHES THE END OF THE STRING...

...GETS REFLECTED AND COMES BACK DOWN...

...BUT IN THE MEAN TIME, MORE WAVES KEEP BEING PRODUCED...

...UNTIL IT BECOMES WHAT LOOKS LIKE A BIG MESS.

SUCH A MESS CAN'T LAST FOR LONG. ALL THOSE WAVES COMBINE INTO JUST A FEW. AT EVERY POINT OF THE STRING, AND AT EVERY MOMENT, THEY INTERFERE WITH ONE ANOTHER CONSTRUCTIVELY OR DESTRUCTIVELY.

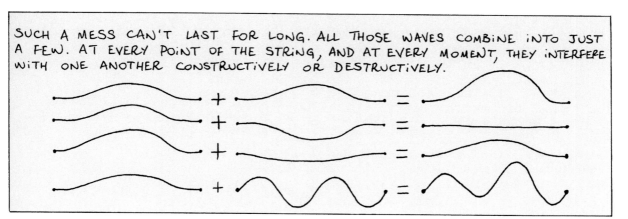

THE RESULT OF ADDING UP ALL THOSE WAVES MUST SATISFY TWO IMPORTANT CONSTRAINTS.

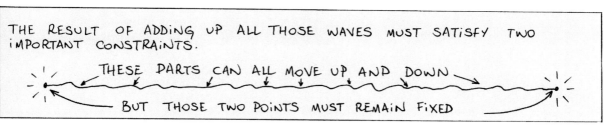

THESE PARTS CAN ALL MOVE UP AND DOWN

BUT THOSE TWO POINTS MUST REMAIN FIXED

AT THE END OF THE DAY, THESE ARE THE WAVES THAT REMAIN:

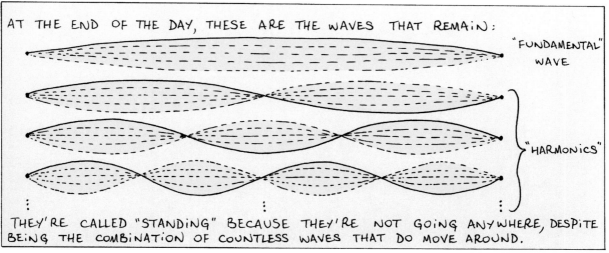

"FUNDAMENTAL" WAVE

"HARMONICS"

THEY'RE CALLED "STANDING" BECAUSE THEY'RE NOT GOING ANYWHERE, DESPITE BEING THE COMBINATION OF COUNTLESS WAVES THAT DO MOVE AROUND.

THE COLOR PHOTOGRAPHY PROCESS THAT GABRIEL LIPPMANN INVENTED IN 1891 MAKES USE OF THE SAME PHENOMENON, BUT WITH LIGHT WAVES INSTEAD OF VIBRATION WAVES ON A STRING.

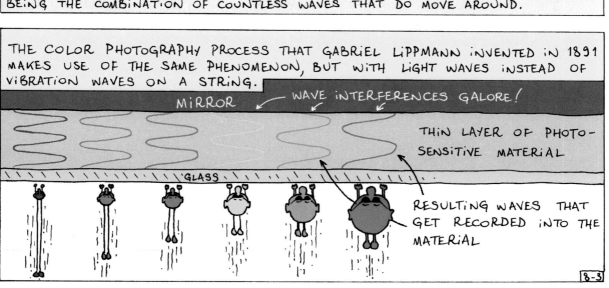

MIRROR — WAVE INTERFERENCES GALORE!

THIN LAYER OF PHOTO-SENSITIVE MATERIAL

GLASS

RESULTING WAVES THAT GET RECORDED INTO THE MATERIAL

IN A WAY, YOU COULD SAY THAT VIEWING A PHOTOGRAPH PRO-DUCED WITH THIS PROCESS YIELDS MORE ACCURATE COLORS THAN OTHER PROCESSES BECAUSE YOU'RE LOOKING AT LIGHT OF THE EXACT SAME WAVELENGTH (PHOTONS WITH THE SAME COLOR & LEG LENGTH) AS THE ORIGINAL SOURCE INSTEAD OF A COMBINATION OF RED, GREEN AND BLUE.

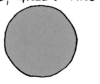

BACK IN 1891 THIS WAS ACTUALLY THE VERY FIRST COLOR PHOTOGRA-PHY PROCESS, WHICH WAS ENOUGH TO GRANT LIPPMANN THE NOBEL PRIZE IN 1908.

HOWEVER, THE PROCESS HAD A FEW MAJOR FLAWS.

?!

IT REQUIRED VERY LONG EXPO-SURE TIMES.

PARFAIT!

NOW STAY VERY STILL FOR 30 MINUTES.

AND THE RESULTING PHOTOGRAPH WAS IMPOSSIBLE TO DUPLICATE.

WAIT, WHAT ??

SO IT NEVER REALLY TOOK OFF, AND INSTEAD WAS SOON REPLACED BY MORE PRACTICAL INVENTIONS SUCH AS THE "AUTOCHROME" PLATE, MODERN FILM, AND OF COURSE DIGITAL PHOTOGRAPHY.

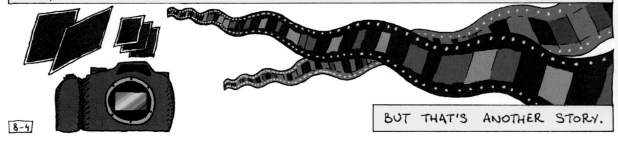

BUT THAT'S ANOTHER STORY.

1909

GUGLIELMO MARCONI & KARL FERDINAND BRAUN

In recognition of their contribution to the development of wireless telegraphy.

YOU LIKELY NEED LESS THAN 10 SECONDS TO NAME 3 UBIQUITOUS TECHNOLOGIES THAT COULDN'T EXIST WITHOUT WIRELESS COMMUNICATION.

WHEN HEINRICH HERTZ, WHO FIRST DISCOVERED THIS FORM OF COMMUNICATION...

... WAS ASKED WHAT COULD BE THE APPLICATIONS OF HIS DISCOVERY, HE SAID:

NOTHING, I GUESS...

SCIENTIFIC GENIUS AND ENGINEERING CREATIVITY DON'T ALWAYS GO HAND IN HAND.

THIS GUY FERDINAND BRAUN, THOUGH, HELPED REFINE THIS TECHNOLOGY AND SAW ITS POTENTIAL...

... BUT NOT AS MUCH AS GUGLIELMO MARCONI, WHOSE BUSINESSMAN SIDE CLEARLY SAW THE OPPORTUNITY.

BUSINESSMAN SIDE

HERTZ WOULD LIKELY HAVE SHARED THE 1909 NOBEL PRIZE, HAD HE NOT PASSED AWAY IN 1894 AT AGE 36.

(YOU CAN'T GET THE PRIZE POSTHUMOUSLY.)

9-1

THE THREE OF THEM FOCUSED MAINLY ON RADIO WAVES. THERE'S NOTHING MAGICAL ABOUT THEM, THEY'RE JUST ANOTHER TYPE OF LIGHT, BUT WITH SHORTER LEGS.

RADIO

AND NOW WE'RE BACK TO THE QUESTION: HOW IS LIGHT CREATED?

REMEMBER WE MENTIONED THAT IT CAN BE CREATED WHEN MICROSCOPIC PARTICLES MOVE BACK AND FORTH REPEATEDLY?

IN THIS CASE, IT'S GOING TO BE ELECTRONS MOVING INSIDE AN ELECTRIC CIRCUIT.

SO LET'S WALK IN HERTZ, BRAUN AND MARCONI'S FOOTSTEPS AND BUILD ONE OF THOSE. WE ARE GOING TO NEED A FEW BUILDING BLOCKS.

ELECTRONS IN A CIRCUIT ARE A LOT LIKE BALLS IN A GIANT PINBALL MACHINE. HAVE YOU PLAYED PINBALL? IT ALL STARTS WITH THAT SPRING-LOADED ROD CALLED "PLUNGER".

SHTOINK   WHEEEE~

FOR OUR CIRCUIT WE'RE ACTUALLY GOING TO NEED AN AUTOMATIC PLUNGER THAT SHOOTS REPEATEDLY FROM A LARGE STOCK OF PINBALLS.

SHTOTOTOTOTOTOTUI!!!!

AT THE OTHER END, WE HAVE THE "DRAIN". THAT'S WHERE PINBALLS GO WHEN YOU LOSE.

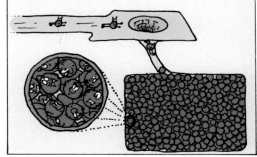

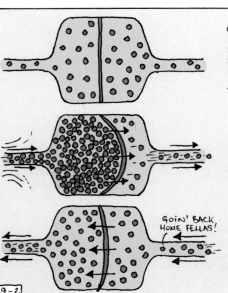

OUR THIRD BUILDING BLOCK IS A SORT OF CYLINDER WITH A RUBBER SEPARATION INSIDE. IT STARTS OFF WITH ROUGHLY THE SAME NUMBER OF BALLS ON BOTH SIDES.

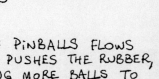

IF A POWERFUL STREAM OF PINBALLS FLOWS INTO IT FROM ONE SIDE, IT PUSHES THE RUBBER, AT THE SAME TIME ALLOWING MORE BALLS TO BE STORED ON ONE SIDE, AND EJECTING SOME OUT AT THE OTHER END.

GOIN' BACK HOME FELLAS!

ONCE THAT STREAM STOPS, THE RUBBER WILL TRY TO GO BACK WHERE IT STARTED, PUSHING OUT THE STASH OF ACCUMULATED BALLS AGAIN, AND ALLOWING SOME BACK IN ON THE OTHER SIDE.

AND THEN THERE'S THE BIG TRAP DOOR. PRETTY STRAIGHTFORWARD:

IT'S INITIALLY CLOSED...

...AND BALLS ACCU-MULATE ON IT...

...UNTIL THERE'S ENOUGH WEIGHT TO FORCE IT OPEN...

...AND THEN IT FLAPS SHUT AGAIN.

FLAP

SHTOK

IF YOU'VE EVER SEEN A STEAMBOAT OUR NEXT BUILDING BLOCK WILL LOOK FAMILIAR: A PADDLE WHEEL. EXCEPT THAT THIS ONE HAS HEAVY WEIGHTS AT THE TIP OF ITS PADDLES.

HEAVY

THE WHEEL IS INI-TIALLY AT REST.

WHEN BALLS ARE BEING SHOT AT IT FROM ONE SIDE, IT STARTS ROTATING, WITH A LOT OF INERTIA.

AND ONCE IT'S GOING THE BALLS CAN FLOW WITH VERY LITTLE RESISTANCE.

BECAUSE OF THE WEIGHTS, IT WILL KEEP ROTATING AND PUSHING BALLS FOR-WARD FOR A WHILE EVEN AFTER THE BALLS STOP ARRIVING WITH ANY SPEED AT ALL.

LASTLY, HERE IS THE PIECE THAT IS ACTUALLY GOING TO PRODUCE THE (SHORT-LEGGED, RADIO) PHOTONS: THE ANTENNA. THIS ONE IS MERELY A COUPLE OF TUBES BENT IN A STRANGE WAY.

CLOSED
OPEN

WHEN AT REST, IT IS FILLED WITH A SMALL NUMBER OF BALLS; ONCE CONNECTED TO AN ACTIVE CIRCUIT, THE BALLS START PERFORMING THEIR LITTLE DANCE.

① ② ③ ④ ⑤

IF YOU SQUINT AND JUST LOOK AT THE TOP PART, IT LOOKS A LOT LIKE A SIMPLE BACK-AND-FORTH OF A BUNCH OF PINBALLS (A.K.A. ELECTRONS), WHICH IN TURNS PRODUCES PHOTONS.

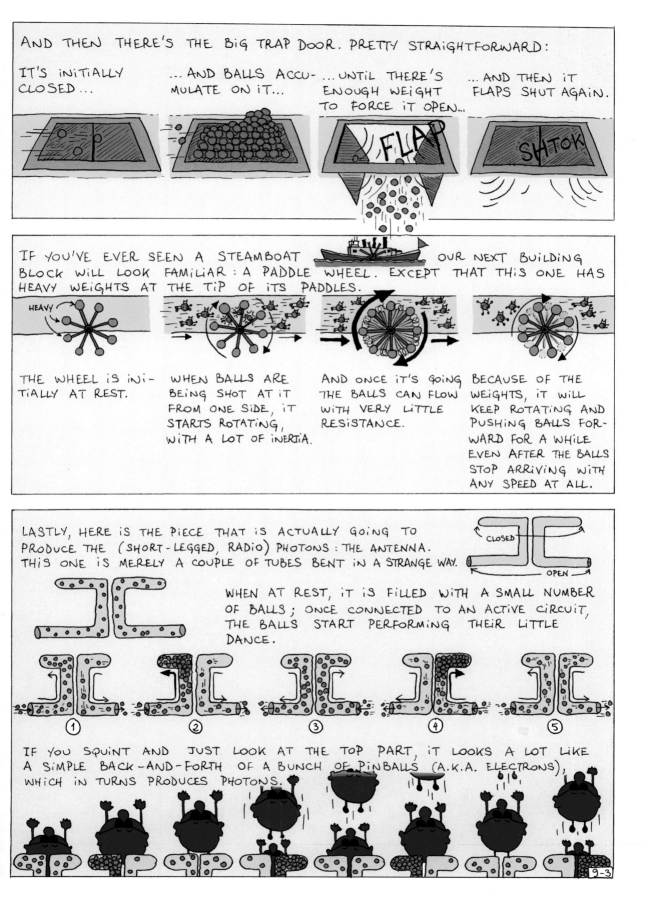

9-3

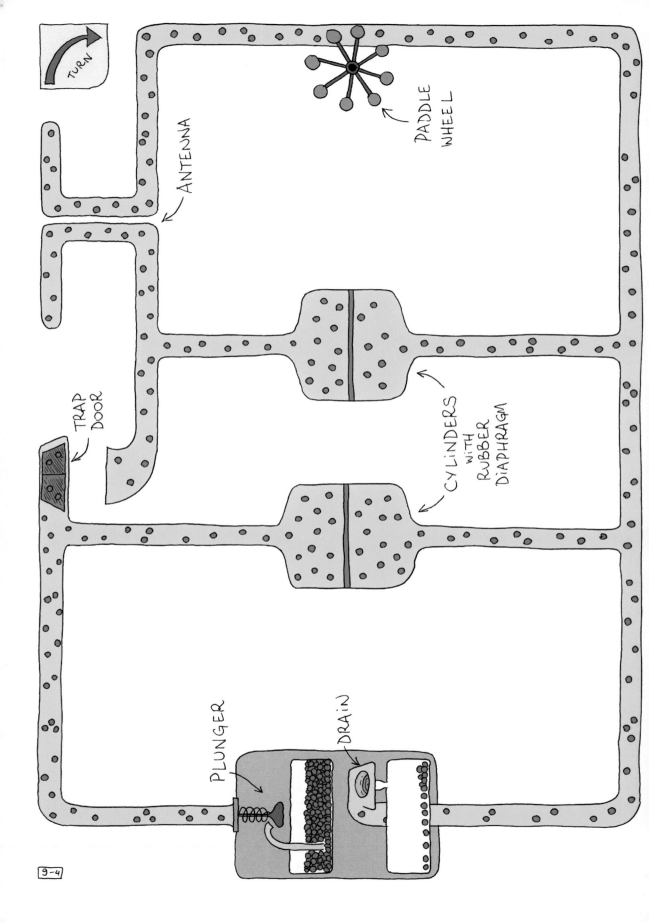

TURN

PADDLE WHEEL

ANTENNA

TRAP DOOR

CYLINDERS WITH RUBBER DIAPHRAGM

PLUNGER

DRAIN

9-4

WATERFALL MOMENT

FULLY STRETCHED

SLOWLY STARTING TO MOVE

SPINNING AT FULL SPEED

STARTING TO PUMP AGAIN

BECAUSE OF INERTIA STILL PUSHING BALLS DOWN

SATURATED

CAN'T PUMP ANYMORE

START HERE

NOW LET'S PUT EVERYTHING TOGETHER AND SEE WHAT HAPPENS.

IF YOU THINK ALL THIS IS GONNA DO IS GIVE YOU A HEADACHE, SKIP IT AND JUST REMEMBER IT SHOOTS OUT SHORT-LEGGED PHOTONS.

STOPPED

STARTS TO SPIN AGAIN, OPPOSITE DIRECTION

PUMPING

9-5

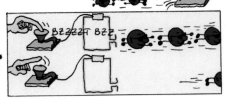

IN PRACTICE, THE FLOW OF PINBALLS (ELECTRONS) THRUST BY THE "PLUNGER" WAS CONTROLLED WITH A SWITCH.

THIS WOULD ALLOW AN OPERATOR TO SEND TEXTUAL MESSAGES IN MORSE CODE (SOUND TRANSMISSION WOULD COME LATER).

 HERTZ WAS AT THE CORE OF THE FUNDAMENTAL DISCOVERY THAT MADE THIS TECHNOLOGY POSSIBLE. TODAY, HIS NAME IS A PHYSICAL UNIT SYNONYM OF "TIMES PER SECOND".

 BRAUN WAS MORE OF AN INVENTOR; HE WAS INSTRUMENTAL IN ENHANCING THE TECHNOLOGY. (HE ALSO HAPPENS TO HAVE INVENTED THE FIRST CATHODE RAY TUBE).

MARCONI REALLY POPULARIZED WIRELESS TELEGRAPHY AND MADE IT A COMMERCIAL SUCCESS. HE HELD 37 PATENTS (BRAUN HAD 2, HERTZ HAD NONE)

◀ THEORY — PRACTICE ▶

THE MAIN ISSUE WITH THIS EARLY RADIO TRANSMITTER THAT WE'VE DESCRIBED (IT'S CALLED A "SPARK GAP TRANSMITTER") IS THAT IT ACTUALLY SPITS OUT A FAIRLY WIDE RANGE OF PHOTONS OF VARIOUS LEG LENGTHS INSTEAD OF BEING WELL DEFINED. THIS MEANS THAT IT INTERFERES WITH OTHER WIRELESS COMMUNICATIONS A BIT TOO MUCH.

NOT ENOUGH ROOM...   ...FOR OTHER COMMUNICATIONS   THIS TYPE OF RADIO TRANSMISSION IS THUS BANNED WORLDWIDE SINCE 1934

NEVERTHELESS, SPARK GAP TRANSMITTERS QUICKLY PROVED EXTREMELY USEFUL, SUCH AS ON 14 APRIL 1912 WHEN IT ALLOWED THE RMS TITANIC TO ALERT NEARBY SHIPS, SAVING 706 PEOPLE.

BUT THAT'S ANOTHER STORY.

	SILLY NAME	REAL NAME	WHAT IT LOOKS LIKE IN REAL LIFE
	PLUNGER & DRAIN	BATTERY	
	CYLINDER WITH RUBBER DIAPHRAGM	CAPACITOR	
	TRAP DOOR	SPARK GAP	
	PADDLE WHEEL	SELF INDUCTOR	
	ANTENNA	ANTENNA	

9-7

PHOTON

1910

JOHANNES DIDERIK VAN DER WAALS

For his work on the equation of state for gases and liquids.

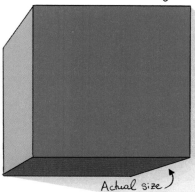

IMAGINE THAT THIS CUBE RISES ABOVE THE PAGE, INTO THE AIR. AND SINCE AIR IS 80% NITROGEN GAS (N_2), LET'S JUST SAY IT IS PURE N_2 AND NOTHING ELSE.

HOW MANY N_2 MOLECULES ARE INSIDE THIS CUBE RIGHT THIS MOMENT?

ASSUMING YOU'RE ON EARTH (HEH), AROUND SEA LEVEL AND AT A COMFY TEMPERATURE, THERE'S ABOUT TEN BILLION BILLIONS. THAT'S ROUGHLY THE NUMBER OF INSECTS ON EARTH.

Actual size →

AT LEAST THAT'S WHAT SOMETHING CALLED THE "LAW OF IDEAL GASES" TELLS US. BUT WAIT A SECOND. HOW COME THIS DOESN'T DEPEND ON WHAT GAS WE ARE TALKING ABOUT?

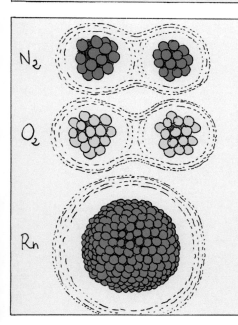

N_2

O_2

Rn

WHAT IF IT WAS O_2 INSTEAD OF N_2? IF THE TEMPERATURE, VOLUME AND PRESSURE ARE THE SAME, ALSO TEN BILLION BILLIONS.

WHAT IF IT WAS RADON? DITTO.

WHAT IF WE INVENT A MOLECULE SO HUGE THAT IT LOOKS LIKE THIS? THE IDEAL GAS LAW SAYS: "I DON'T CARE. TEN BILLION BILLIONS."

WELL THAT CAN'T BE RIGHT. THE CUBE UP THERE CAN'T EVEN FIT TEN OF THESE.

To draw cartoons about physics you really gotta like drawing tiny circles. →

↓ Actual size

10-1

THAT IS LIKELY THE SORT OF "THOUGHT EXPERIMENTS" THAT JOHANNES DIDERIK VAN DER WAALS, DECIDEDLY A THEORETICAL PHYSICIST, LIKED DOING. RIGHT AROUND 1873.

THE ELDEST OF 10 CHILDREN IN A WORKING CLASS DUTCH FAMILY, HE HAD TO LEAVE SCHOOL AT AGE 15. ONE CAN ONLY BE IMPRESSED AT HOW HE OVERCAME THE SOCIAL DIVIDES IN 19th CENTURY EUROPE.

ANOTHER THOUGHT THAT MUST HAVE BEEN ON VAN DER WAALS'S MIND WAS THE TEMPERATURES AT WHICH VARIOUS LIQUIDS EVAPORATE: WHY ARE THEY ALL SO DRASTICALLY DIFFERENT?

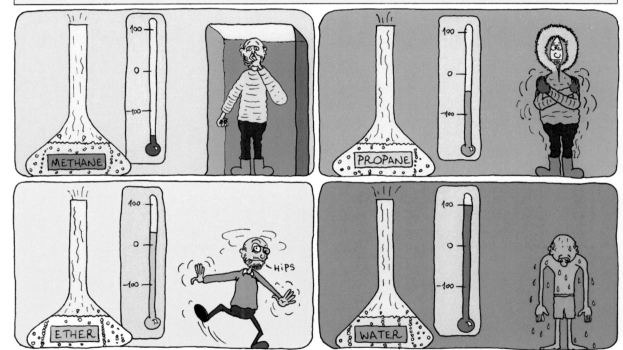

THE "IDEAL GAS LAW", WHICH WAS STILL THE BLEEDING EDGE AT THE TIME, COULDN'T EXPLAIN THAT EITHER.

SO VAN DER WAALS WENT BACK TO HIS DRAWING (BLACK) BOARD AND DESIGNED TWO ENHANCEMENTS TO THAT LAW: ONE ABOUT SIZE AND ONE ABOUT ATTRACTION (BUT DON'T GET YOUR HOPES UP).

10-2

FIRST, SUPPOSE YOU HAVE A TINY GARDEN, ABOUT THE SIZE OF A TWO-PERSON BED, AND YOU DECIDE TO START BREEDING SOME ANIMALS. SAY, 10 OF THEM TO START. DEPENDING ON WHICH SPECIES YOU PICK, THOSE POOR FELLOWS MAY OR MAY NOT LIKE IT VERY MUCH.

MMOOO

SAME THING WITH MOLECULES: FOR A GIVEN SPACE, BIG ONES WILL HAVE LESS SPACE TO MOVE AROUND THAN SMALL ONES.

WELL, DUH !!

REMEMBER THAT THIS WAS 1873, WHEN THE VERY CONCEPT OF MOLECULE WAS STILL A THEORY. IN FACT, THERE WERE STRONG OPPONENTS TO THE IDEA OF ATOMS. THINK ABOUT THIS: YOU HAVE A GLASS OF WATER, FROM WHICH YOU CAN POUR OUT ANY AMOUNT WITH INFINITE PRECISION. YOU POUR OUT HALF. THEN YOU POUR OUT HALF OF WHAT REMAINS. YOU KEEP DOING THAT, EACH TIME REDUCING WHAT YOU HAVE BY HALF. CAN YOU KEEP DOING THIS FOREVER? OR WILL THERE BE A POINT WHEN THE RESULT OF THE HALVING IS NO LONGER WATER?

ANOTHER "THOUGHT EXPERIMENT". GREAT...

$\Rightarrow \infty$

STOP

TODAY WE KNOW THE LATTER TO BE TRUE, BUT IT'S BY NO MEANS OBVIOUS, AND SOME PHYSICISTS BACK THEN WERE CONVINCED OF THE CONTRARY.

JOHANNES! COME MOP THIS UP RIGHT NOW!

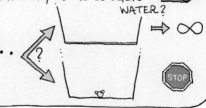

WHAT WE ALSO KNOW TODAY IS THAT TEMPERATURE IS A SYNONYM FOR HOW FAST MOLECULES MOVE AROUND...

...AND THAT PRESSURE IS A SYNONYM FOR THE CONSTANT COLLISIONS OF MOLECULES AGAINST A SURFACE.

OUCH
OW
OWW
HEY LOOK WHERE YA GOIN'!

A GAS WILL BEHAVE LIKE AN "IDEAL GAS" IF IT'S EASY FOR ITS MOLECULES TO IGNORE ONE ANOTHER: SMALLER MOLECULES, LOW PRESSURE (FEWER COLLISIONS) AND HIGH TEMPERATURE (MOLECULES ZOOM PAST ONE ANOTHER WITHOUT INTERACTING).

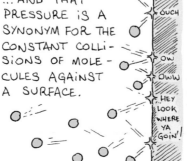

THE SECOND OF VAN DER WAALS'S ENHANCEMENTS RECOGNIZES THAT MOLECULES HAVE A CERTAIN DEGREE OF ATTRACTION TO ONE ANOTHER. TAKE FOR INSTANCE WATER AND OXYGEN.

IT TURNS OUT THAT ELECTRONS ARE A LOT MORE ATTRACTED TO AN O ATOM THAN TO AN H. SO IN H_2O THE ELECTRON "CLOUD" IS LOPSIDED. IN O_2 HOWEVER, NO SUCH BLANKET PULLING.

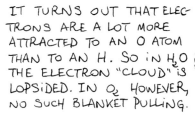
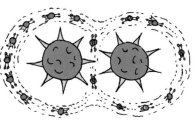

ONE SIDE OF THE H_2O MOLECULE ENDS UP BEING "MORE NEGATIVE" (MORE ELECTRONS — REMEMBER, THOSE ARE NEGATIVELY CHARGED) AND THE OTHER SIDES "MORE POSITIVE", WHEREAS ELECTRIC CHARGE REMAINS PRETTY WELL DISTRIBUTED OVER O_2.

THIS MEANS THAT BY ALIGNING PROPERLY, TWO WATER MOLECULES CAN ACTUALLY BE QUITE ATTRACTED TO EACH OTHER.

BETWEEN THE O_2s, MEH.

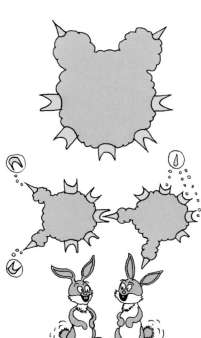

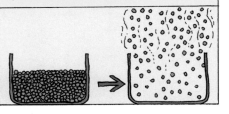

IF YOU WANT TO VAPORIZE A LIQUID, YOU NEED TO PULL THE MOLECULES APART.

SO WHAT THIS ALL MEANS IS THAT IF TWO WATER MOLECULES ARE HARDER TO SEPARATE, YOU NEED A LOT MORE ENERGY TO PULL THEM APART (HENCE THE MUCH HIGHER BOILING POINT).

NOOOOOOOO GEORGETTE!!

MEH

MEH

IT IS THIS SAME PROPERTY OF "ELECTRICAL LOPSIDEDNESS" THAT ALLOWS WATER TO DISSOLVE MORE SUBSTANCES THAN ANY OTHER LIQUID.

BUT THAT'S ANOTHER STORY.

10-4

Afterword

- The history of physics Nobel prize winners is mostly a history of white males, especially early on. I'm afraid there's not much I can do about that.

- Obviously, this book is full of metaphors. Real electrons don't have legs.

- Scientific accuracy is extremely important to me, but I still had to make a few simplifications. If this book describes (and oversimplifies) a phenomenon that you then feel compelled to look up, it will have accomplished part of its purpose: enticing curiosity.

Index

Face

Henri Becquerel
Physics Nobel 1903

Ferdinand Braun
Physics Nobel 1909

Marie Curie
Physics Nobel 1903

Pierre Curie
Physics Nobel 1903

Albert Einstein
Physics Nobel 1921

Electron (a.k.a. cathode ray, a.k.a. β particle)
Negatively charged

Alexander Fleming
Discovered Penicillin

γ Photon
Longest legs, highest energy

Heinrich Hertz
Physicist

Infrared Photon
Short legs, low energy

Philipp Lenard
Physics Nobel 1905

Gabriel Lippmann
Physics Nobel 1908

Hendrik Lorentz
Physics Nobel 1902

Guglielmo Marconi
Physics Nobel 1909

Albert Michelson
Physics Nobel 1907

Neutron
Electrically neutral, part of the nucleus

book

Noble Gas

Doesn't like to mingle

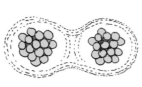

Oxygen

Symmetrical molecule

Proton

Positively charged, part of the nucleus

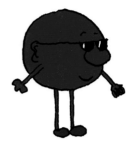

Radio Photon

Very short legs, very low energy

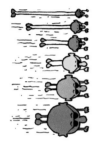

Rainbow Photon

Colorful part of the photon spectrum

William Ramsay

Chemistry Nobel 1904

Lord Rayleigh

Physics Nobel 1904

Wilhelm Röntgen

Physics Nobel 1901

Percy Spencer

Invented the Microwave oven

Joseph John Thomson

Physics Nobel 1906

Ultraviolet Photon

Long legs, high energy

Johannes Van der Waals

Physics Nobel 1910

Andy Warhol

Artist

Water

Very lopsided molecule

X-ray photon

Very long legs, penetrates living tissues

Pieter Zeeman

Physics Nobel 1902

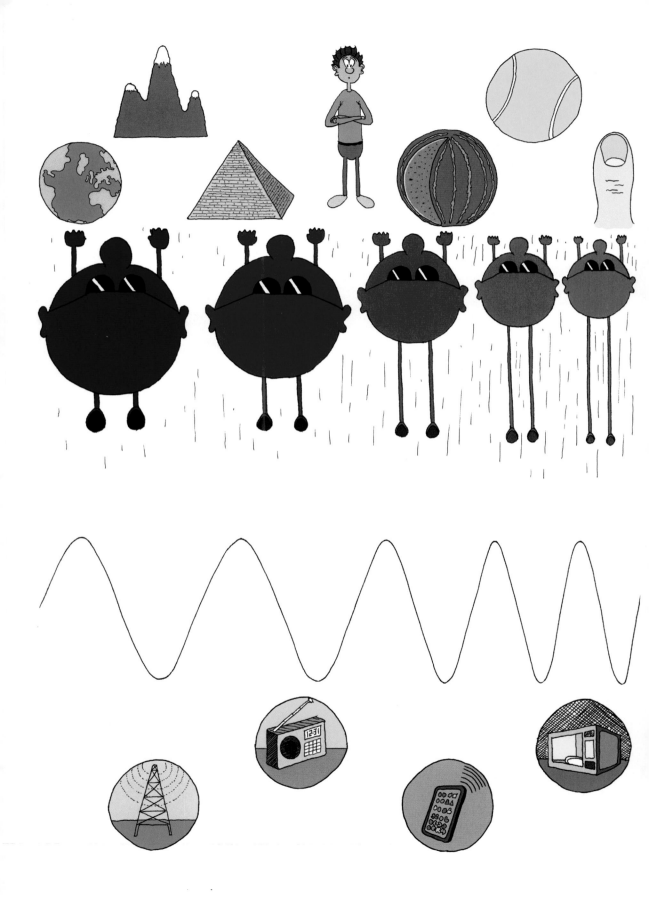

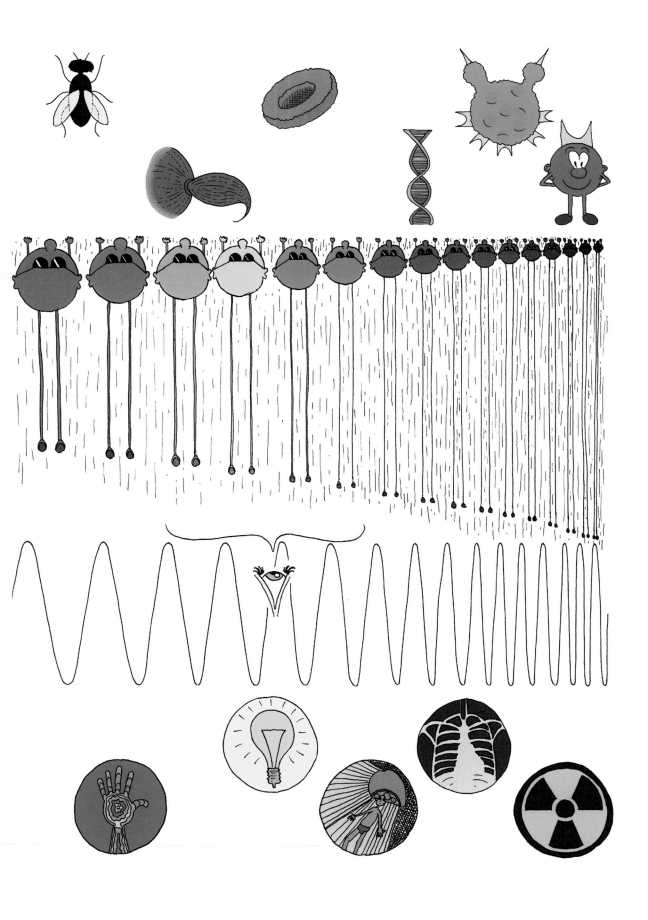

Thank you for reading!

By the same author:

In English

2018

2012

Future works

The Seven Deadly Sins of Tinder (est. 2020)

Goomics: Google's corporate culture revealed through internal comics
Vol. 2, 2015 — 2020: Disillusionment

In French

2012

2006

2006